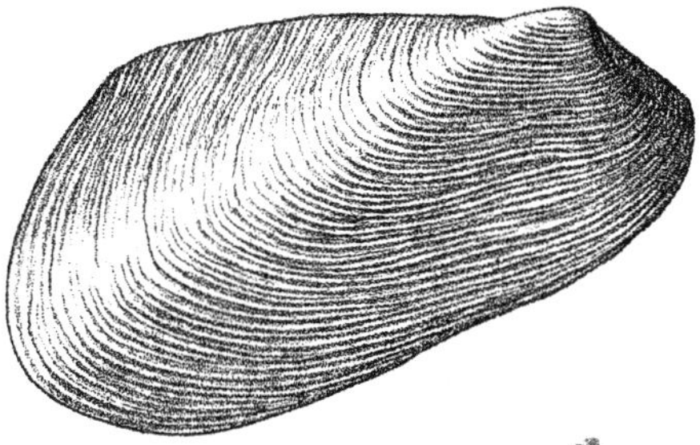

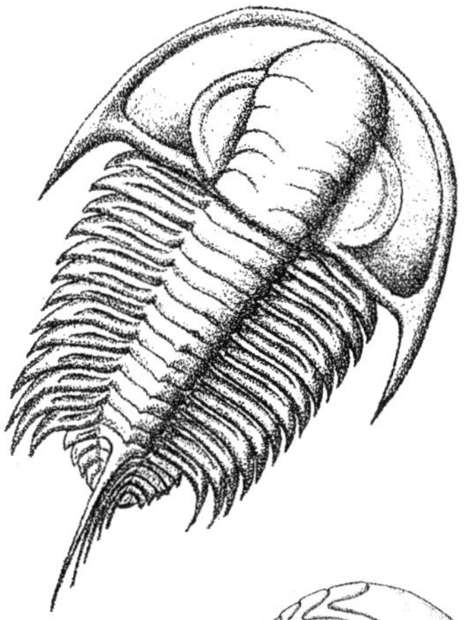

Fossils and "Deep Time" in New York

THE UNIVERSITY OF THE STATE OF NEW YORK

Regents of The University

Fossils and "Deep Time" in New York

by

Ed Landing

New York State Museum
Educational Leaflet No. 35

2004

THE UNIVERSITY OF THE STATE OF NEW YORK
THE STATE EDUCATION DEPARTMENT

Published 2004

Printed in the United States of America

Copies may be ordered from:

Publication Sales
3140 CEC
New York State Museum
Albany, New York 12230
Phone: (518) 402-5344
FAX: (518) 474-2033
Web address: http://www.nym.nysed.gov./publications,html

Library of Congress Catalog Card Number: 2004102399

ISBN: 1-55557-216-2
ISSN: 0735-4401

FOSSILS and "DEEP TIME" in NEW YORK

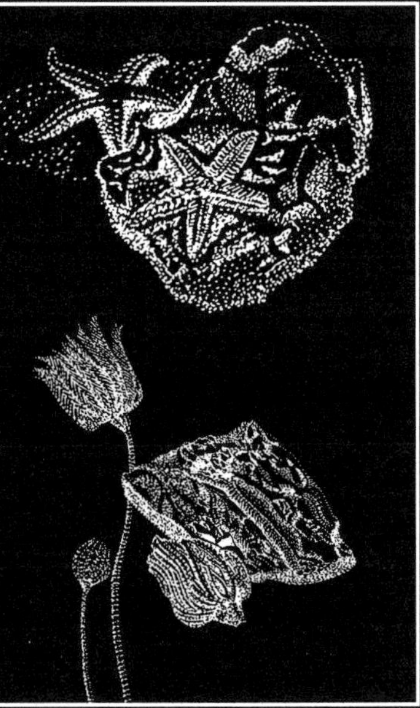

"The past history of the living world . . . a history involving periods of time before which our mere human records sink into utter insignificance"
–Thomas Henry Huxley, 1862, from *On The Origin of Species*

By Dr. Ed Landing
State Paleontologist
New York State Museum

INTRODUCTION

Paleontology, the study of ancient life, illuminates the depths of "deep time"—the inconceivably long interval of life and earth history that preceded and overlapped with early human experience. Paleontology is not only a subdiscipline of geology, but is a major cultural activity that has been shaped by and has also changed humanity's understanding of the earth and its workings. It is trans-national in scope as its research techniques, focus, and technical vocabulary allow unfettered communication between workers in widely separated and distinct cultures. With an advanced degree, a paleontologist can undertake a career in college and university teaching or museum research, collection conservation, and exhibits work. Paleontologists are also involved in the search for oil, gas, coal, or other natural resources in private industry or in state or federal geological surveys. However, the real satisfactions to be taken from paleontology are the insights it provides into "deep time." These insights take place while reading the tales told by the rocks and fossils in building and field stones or walking along a beach or river and seeing how modern organisms and sediments can serve as guides to understanding ancient fossils and rocks.

PALEONTOLOGY IN HUMAN HISTORY

Mankind's interest in the remains of ancient life is relatively ancient. Complex, natural markings in rock, whether symmetrical, coiled, or seemingly life-like, early attracted humanity's attention. For example, the tightly coiled "rams' horns" in early Celtic and Norse art and jewelry were inspired by ca. 100 million year-old ammonite shells (produced by relatives of squids) common in the rocks of central and western Europe. In the early 1970s, archeologists found evidence of the oldest fossil collecting in North America, with the recognition that a large trilobite (an extinct marine arthropod) only found in western British Columbia had been carried to a Shuswap site some 450 kilometers to the west.

Aristotle and Leonardo da Vinci were among the few early western philosophers to conclude that many of these markings were the remains of ancient life. In Medieval Europe, however, a less empirically rational approach was taken to questions involving natural history, and all answers were thought to lie in the writings of classical authors and in the Bible. In addition, scientific inquiry based on field or laboratory study did not develop because of the intellectual distortions associated with the mysticism of alchemy. Fossils were generally regarded as objects created by God, Satan, or mysterious "vital" forces" to confuse humanity. At least two theologically inspired ideas about fossils and earth history were current which persist today in the recurrent court challenges (which only take place in the U.S.) to the teaching of evolution in public schools. The first view was that the earth is relatively young, and as the English Archbishop William Paley concluded in 1802, was created at 9:00 AM on October 26, 4004 B.C. The second was that fossils were organisms that perished in the reshaping of the world during the chaos of Noah's Flood.

There is no scientific data supporting either of these beliefs. Indeed, it should be emphasized that systematic studies of earth history and paleontology began in the late 18th and early 19th century, and were, in particular, furthered by the work of British and French clergyman and "natural philosophers" trained in church-run universities. These early workers saw themselves not in opposition to western religion, but as helpers in the illumination of a slow and grand unfolding of God's plan that they believed to be outlined in *Genesis*. They also saw the need for an accurate understanding of the structure of the earth in the search for the iron, coal, and other materials needed by the Industrial Revolution. By the mid-1800s, considerable data showed that the earth and, likely, other planets of the solar system are very old and that different kinds of animals and plants made their appearance, evolved, and disappeared at different times in the geologic past.

The evidence provided by paleontology, other disciplines of geology, and astronomy that the earth and universe are almost inconceivably ancient and evolved through "deep time" is complemented by new analytical procedures. In particular, geochronology, the dating of earth materials, applies the procedures developed by the nuclear industry to provide quite precise absolute ages on wood and other biologic materials (up to 50,000 years in age) and on volcanic ashes and minerals in other rocks (these techniques can potentially date materials up to tens of billion years old). Procedures that give dates based on several different uranium-lead isotopes are typically applied to individual ancient volcanic ashes—the ages determined by each technique are used to verify one single sample's age. Two things should be remembered: 1) Scientists are very competitive, and would be the first to publicize results that might disprove the techniques and results of their colleagues. Consequently, if the evidence was there, geochronologists would be the first to support the "young earth" speculation of creationists. 2) If a directed creation as claimed by Biblical literalists took place, why did the Creator leave all this evidence for the ancient age of the earth, solar system, and universe? Religious groups have dealt differently with the empirical and testable evidence supplied by the modern evolutionary synthesis for the development of the earth and life. For example, the Vatican Councils of 1950 and 1964 concluded that the evolution of life and of mankind were consistent with the Bible and Catholic theology, while Southern Baptist-dominated advocacy groups regularly challenge the teaching of evolution in public schools.

FOSSILS AND RECONSTRUCTING "DEEP TIME"

In order to appreciate the use of fossils in reconstructing earth history, it is necessary to understand two fundamental principles. The first, the Law of Superposition, states that each layer of rock (and its fossils) is younger than the one below it in an undeformed succession of sedimentary rocks. The Law of Superposition is locally not applicable

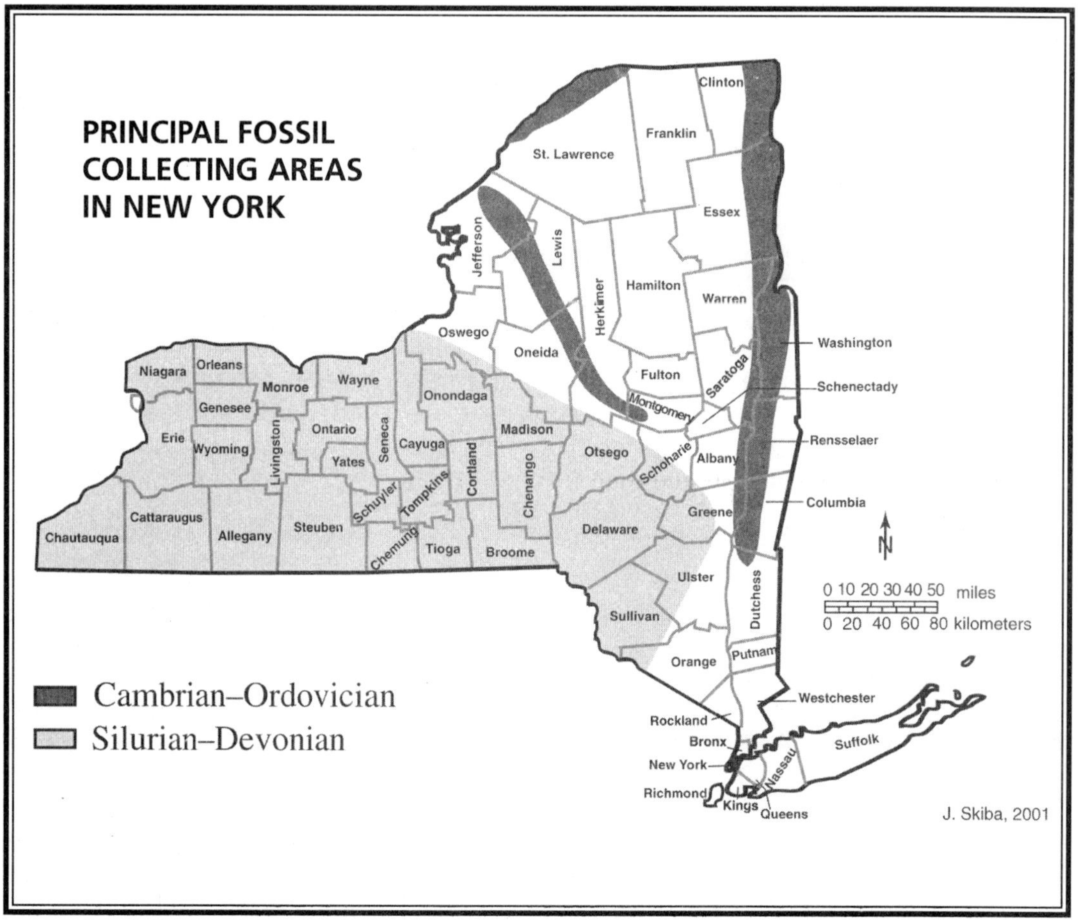

in the deformed rocks of mountain belts, where movements on faults (surfaces along which rocks have been broken and moved) and where intense folding of rocks deep in the earth can bring older rocks over younger rocks. However, these local exceptions to the Law of Superposition are evident when the rock and fossil sequence from a faulted or overturned succession is compared with other successions in the mountain belt and with undeformed successions outside of the mountain belt.

The second is the Law of Biotic Succession—the tested principal that fossils succeed each other in rock successions in a definite and determinable order, and that each interval of time can be recognized by distinctive fossils. The Law of Biotic Succession was recognized long before Charles Darwin's 1859 proposal of his model of evolutionary theory and the adoption of evolutionary theory as the unifying principle of biology and paleontology. Biotic Succession was demonstrated by geologists in the late 1700s and early 1800s, who noted that fossils appear at a particular level in rock successions, persist through a vertical interval of rock, and then disappear, never to reappear. Thus, the vertical distribution of fossils in rocks could be used to distinguish roughly equivalent relative time intervals wherever these fossils were found. This practice of relative time correlation using fossils, or biostratigraphy, was important in prospecting for coal and iron deposits, as these resources are found in rocks of particular ages in western Europe and eastern North America. The upward disappearance of fossils was early equated with extinction, and marked an important divergence from then-current theological thought—in which God's creation was thought to be perfect, and extinction was impossible. The Law of Biotic Succession supports the modern evolutionary synthesis, in which origin through evolution and subsequent disappearance, never to appear again, through extinction limit the temporal range of all organisms. A number of organisms that have persisted virtually unchanged for millions of years are called "living fossils." The coelocanth, a cousin of the lungfish known from the Indian Ocean; the pearly nautilus, a shelled squid-like mollusk from the same region; the horseshoe crab; and the insects known as silverfish are all "living fossils." They differ little from fossil ancestors that lived hundreds of millions of years ago.

Particularly useful in biostratigraphy are "guide" or "index fossils." These are relatively short-lived forms (i.e., they have a limited vertical range in rock successions following their initial evolution and subsequent extinction) which were abundant and geographically widespread, tend to be easily identified, and allow correlation of rocks between distant areas. Individual "index fossils" are less relied on for correlation. Rather, associations of a number of fossils ("assemblages") are customarily used to divide rock successions into relative time intervals called "biotic zones," and are relied on to correlate by biotic zones between regions.

FOSSIL PRESERVATION

The soft and hard parts of most animals and plants are generally completely recycled after their death as food and nutrients by predators, scavengers, decay organisms, and non-biologic processes (e.g., waves, currents). Only a few remains survive the organisms that produced them. When preserved, three types of fossils are recognized. 1) Body fossils (bones, teeth, shells, leaves, petrified wood, body imprints in once-soft sediments, rare frozen and mummified remains) are the remains of an organism. 2) Trace fossils are fossils of organism behavior, and include tracks, trails, burrows in once-soft sediment, root casts, and borings in hard rock or shells. 3) Chemical fossils are the chemical products derived from organisms (e.g., coal, petroleum, amber [fossil plant resin], and other organic chemicals).

The process of fossilization begins with burial of biologic remains in sediments. Subsequent history then leads to the production of sedimentary rocks from sediments. The important processes include compaction (with pressure and the loss of water, the fine grains of muds or muddy sandstones are literally packed together to form rock) and cementation (mineral cements as silica, lime, or other minerals are precipitated onto sediment grains). In diagenesis (the process by which sediments become rock), gravel and pebbles, sand, silt, mud, shell fragments and lime mud, and peat are changed into conglomerate, sandstone, siltstone, shale, limestone, and coal, respectively. Most fossils are found in sediments or sedimentary rocks. Only locally are fossils found in metamorphic rocks (slate, marble) or erupted igneous rocks (lava, volcanic ashes). Aquatic plants and animals (especially forms with hard parts) are more likely to be preserved than land organisms because the water-dwellers tend to be quickly buried in sediment after death. In addition, the remains of land organisms are more likely to be broken down by organisms (scavengers, bacteria, fungi, trampling by large animals) and physical processes (drying, wetting, freeze-thaw, and ultraviolet rays from sunlight decompose even resistant bones and teeth).

Much of New York State is mantled by sand and clay less than 20,000 years old. Because most of this sediment is composed of bed rock debris eroded and transported by continental glaciers, fossils normally do not occur in these late "Ice Age" deposits. Spectacular finds of mastodons, caribou, and birds (including condors) are regularly found in bog and shallow-lake deposits that form part of this late "Ice Age" cover. Beneath these surficial sediments, however, are vastly older marine sedimentary and weakly metamorphosed rocks across much of New York State. These fossiliferous rocks range from ca. 520 million years ago to about 350 million years ago. The much older (1.1+ billion year-old) rocks of the Adirondack Mountains and Hudson Highlands are strongly metamorphosed and almost entirely unfossiliferous. The generalized geologic map indicates the best areas for fossil collecting in New York State.

BRINGING FOSSILS TO LIFE

Fossils should be regarded as once-living organisms that had coloration, behaviors, and ecologic roles similar to their living relatives. Experience helps form the best paleontologist. Personal and second-hand experience (through in-depth reading, videos, etc.) with modern organisms from a wide range of habitats help illuminate the fossils and ancient marine environments of New York. The paleontologist's education should include trips to the open sea shore and *Spartina* grass flats of Cape Cod or Long Island, the mud flats of the lower Hudson River tidal estuary, the barrier islands and lagoons of the Carolinas, and to Florida coral reefs and mangrove swamps. Guidebooks to seashells and other marine animals of the Atlantic and Pacific coasts of the U.S. (e.g., the Audobon, Peterson, Golden Books, and Simon & Schuster's guides) help instruct in the methods that modern (and thus fossil) organisms are classified and identified. These guides outline the very different marine habitats that can occur along and seaward of a shoreline, and are relevant to understanding ancient shorelines. A visit to the library or science section of an independent or "big box" bookstore can yield a treasure trove of books on the biology and ecology of modern organisms, evolution, geology (note in particular the *Roadside Geology of New York* and other states from Mountain Press), and paleontology. Publications available from the New York State Museum in Albany, Rochester Museum of Science, and Museum of the Earth in Ithaca include regional geology summaries, guides to fossil identification, and "do-it-yourself" field trip guides to various areas of New York.

Over the past thirty years paleontology has come "alive" as a discipline. Paleontology not only relates life to profound changes that have taken place on earth, but it also integrates life as a forcing mechanism that has changed earth processes. For example, the role of repeated mass extinctions by periodic asteroid/comet impacts in shaping the changing composition of marine and land communities has received considerable discussion and argument pro and con. Similarly, the appearance of important tropical limestone deposits and coral reefs in New York no longer remains mysterious. The eastern margin of North America was indeed tropical and located south of the equator in the Middle Paleozoic and only later moved north as a result of plate tectonic processes.

As a further example of the insights supplied by modern paleontology, research in the late 1990s at the N.Y.S. Museum reconstructed 80 million years of ancient global climate changes from rocks deposited offshore of ancient tropical New York. In particular, the "Slate Belt" from Washington County to Dutchess County in easternmost New York shows a regular alternation of black- and dominantly green-colored slates. Abundant, large trace fossils in the green slate intervals show that many organisms lived in the muddy, deep-water, ancient continental slope environments recorded by the green slates. However, the black slate alternations have rare and generally small trace fossils. These and other data (e.g., the black slates are rich in organic matter) suggest that the black slates record very poorly oxygenated bottom waters on the continental slope. This situation is consistent with a model that predicts limited storm activity, limited sources of dense and cold polar waters that sink and oxygenate the deep ocean, and consequent limited deep-water circulation during global warming conditions. All of these conditions would be reversed during the "ice house" conditions of the green slates. Thus, rock colors and organism abundance in 520–440 million year-old slates of easternmost New York may predict a future of poorly oxygenated, organism-poor, deep-sea environments with continued global warming.

An interesting contribution to paleontology was made in the 1960s by the late Dr. John Wells of Colgate University. Wells collected exquisitely preserved Middle Devonian (ca. 380 million year-old) corals and brachiopods (clam-like marine animals) from the Finger Lakes region of central New York. He believed that the very fine ridges on these animals' shells were daily growth lines, and noted that they were grouped into apparent monthly and annual bands. His conclusion was that there were 385 "daily" ridges in an annual band! The Middle Devonian year had 385 days—a number predicted by physicists who calculated a slowing of the earth's rotation as a result of tidal friction with the moon.

CLASSIFYING AND IDENTIFYING FOSSILS

Classification is the systematic grouping of "things" (e.g., objects, materials, organisms, ideas, etc.) into categories by specified criteria. Thus, rocks are grouped by their mode of formation (i.e, igneous, sedimentary, metamorphic) and minerals by their composition (elemental metals, oxides, silicates, carbonates, etc.).

Biological classification attempts to group organisms by their degree of relatedness, and thus is phylogenetic (pronounced fie-*loe*-jeh-*neh*-tic) as it attempts to reflect evolutionary descent. Biologic classification involves a hierarchy (an arrangement into a graded series) of organisms. For animals, the hierarchy, from most broadly to most closely related, includes the categories kingdom, phylum (fie-*lum*), class, order, family, genus, species. The only category that is "real" for an individual living plant or an animal is the species—the sum of all those populations of related individuals that interbreed or are capable of interbreeding in nature. Species typically do not interbreed in nature because of physiological (genetic and chromosomal) differences that preclude viable offspring and behavioral differences during reproduction. The next higher category above the species, the genus, includes a number of closely related species. Some of these species, as mallard and pintail ducks, may actually interbreed if brought together in an "unnatural" setting, like a cage. The binomial (two-word) combination of a genus and species name, both of which must be written in italics (e.g., *Tyrannosaurus rex*, the giant meat-eating dinosaur, or *Homo sapiens*, all humans), uniquely designates a species.

Higher in the hierarchy, the family includes related genera, and the order, one or more closely related families. Far fewer classes (groups such as birds, reptiles, bony fish, trilobites) exist than orders. Only twenty or so phyla (fie-*lah*; plural of phylum) are recognized. Phyla represent the major founding stocks and structural plans that originated with the diversification of animals at about the beginning of the Cambrian Period (ca. 543 million years ago).

Once a classification scheme is established, the criteria that define a particular taxon (i.e., a particular species, genus, family, etc.) are used in identification of an organism. "Identification" is fitting the organism into a classification scheme. For much biologic and paleontologic work, identification and the development of classification schemes may be primarily morphologic, and may depend on hard and soft part anatomy and morphology, possibly supplemented by genetic and biochemical evidence in living forms. The following examples show how Neanderthal humans and a trilobite common in central New York (in parentheses) are classified:

Kingdom	Animalia	(Animalia)
Phylum	Chordata	(Arthropoda)
Class	Mammalia	(Trilobita)
Order	Primates	(Phacopida)
Family	Hominidae	(Phacopidae)
Genus	*Homo*	(*Phacops*)
Species	*sapiens*	(*rana*)
Subspecies	*neanderthalensis*	(*rana*)

COMMON NEW YORK FOSSILS

The great areal extent of rocks with abundant fossils and the dedicated work of people like James Hall, the second N.Y.S. Paleontologist, meant that paleontology developed as an important basic and applied science in 19th century America in New York. Since the mid-1830s, over 15,000 types of fossils have been described, many for the first time, from New York. A short summary of all of New York State's fossils and where they are found is obviously impossible. Only a brief survey of the common groups of fossils found in New York State rocks is presented below. This list focuses on invertebrate fossils (organisms that lack a dorsal nerve cord and "backbone"), and not the Devonian-age and younger fish and plants; dinosaur tracks of Rockland County; and widespread "Ice Age" fish, birds, and mammals of the state.

Phylum Porifera

(por-*if*-uhr-ah; "pore-bearers")
Sponges are vase-like, saucer-shaped, or barrel-like animals that are common as sessile, attached forms in marine waters, and rare in fresh water. A few marine types occupy tiny borings that they produce in shells and corals. Some of them secrete mineralized skeletons composed of calcium carbonate (lime) or opal (a form of hydrated silica. *Hydnoceras* in the illustrations had an opaline skeleton. Bath sponges have a flexible framework composed of spongin, a protein.

Class Stromatoporoidea

(stroh-mah-*tah*-pore-*oi*-dee-ah; "layered pore bearers")
These extinct marine forms differ from other sponges in that they produced hard, finely layered calcium carbonate build-ups that, at times, resemble coral heads. Fine pores can be seen with a hand lens on the surface and in broken sections of stromatoporoids. They lived in clear, shallow seas and were important reef builders in Ordovician—Devonian seas in New York.

Phylum Cnidaria

(nih-*dare*-eeh-uh "forms with stinging cells")
Marine and fresh-water forms, many of which are colonial. They have a simple sack-like body construction with a mouth surrounded by tentacles with stinging cells (used to capture prey) and a deep gut. Sea anemones, jellyfish, sea pens, corals, and hydras are cnidarians.

Class Scyphozoa

(sky-foh-*zoh*-ah; "cup animals")
Jellyfish are an exclusively marine, non-colonial group of cnidarians. They are generally free-swimming as adults, lack hard parts, and are seldom preserved as fossils. An extinct group termed conulariids (*kahn*-you-*lair*-ids, see the illustrations) are known from four-sided, phosphatic, cone-like fossils. Conulariids are considered sessile jellyfish by many paleontologists.

Class Anthozoa

(*anne*-tho-*zoh*-ah; "flower animals;" named for the tentacles that radiate away from the mouth)
Corals secrete hard, calcareous skeletons, and are the second most abundant fossil group, after brachiopods (discussed below), in New York, where they represent two major orders. 1) Rugose corals include solitary forms with horn-shaped skeletons ("horn corals," see *Zaphrentis* in the illustrations) and colonial types that constructed dome- and head-shaped build-ups. 2) Tabulate corals are colonial forms with tiny coralites (the small calcareous tubes that the polyps, or individual coral animals, lived in). *Favosites* is a Silurian and Devonian tabulate coral often called a "honey-comb coral" for the appearance of its colonies. Many reefs

in Ordovician–Devonian rocks in New York are constructed primarily of corals.

Phylum Bryozoa

(*bry*-oh-*zoh*-ah; "moss-animals," named for the moss-like form of some modern colonies)
Bryozoans (see illustrations) are colonial marine and (rare) fresh-water animals. The zooids (individual animals), which are generally smaller than a coral polyp, work together to build stick-, fan-, and dome-like build-ups that are frequently calcareous. They were important contributors to reefs, and are most common in limestone and calcareous shale.

Phylum Brachiopoda

(*brah*-kee-oh-*poe*-dah; "arm foot")
Brachiopods (see *Spirifer* in the illustrations), which resemble clams, are the most abundant fossils in New York. These marine animals always have two valves (shells) that are opened and closed by muscles. The shells lie above and below the animal, and are bilaterally symmetrical. By comparison, clam shells lie on the right and left sides of the animal, and are not bilaterally symmetrical (see *Glyptodesma* in the illustrations). Most brachiopods are fixed to the bottom throughout life. An inability to escape predators may explain the decline in brachiopods and the diversification of mobile clams with the appearance of jawed fish in the Devonian Period.

"Worms"

Several phyla of elongate, soft-bodied animals are commonly lumped as "worms" (see illustrations). They are uncommon as body fossils, but trails, burrows, or excreta (coprolites) produced by unknown, unpreserved "worms" are common in marine rocks in New York State. Ancient worms probably similar to the modern sand worm of the Atlantic coast had complex jaws. These microscopic jaw parts (called scolecodonts) are composed of resistant organic compounds (polysaccarides), and can be dissolved out of many limestone samples with acid.

Phylum Mollusca

(moe-*luss*-kah; "soft bodied")
This large group of marine, fresh-water, and land animals is characterized by a muscular foot and, generally, a dorsal calcareous shell or shells. Mollusks include snails and slugs, clams, squids, octopi, chitons, tusk shells, and a number of other living and extinct groups.

Class Bivalvia

(by-valve-a; "two shells")
Clams (including mussels, oysters, and scallops) are marine to fresh-water animals. The soft body, with an poorly defined head, gills, and a typically well developed foot, is enclosed in two shells. These shells are commonly mirror images (i.e., right and left) of each other (see *Glyptodesma* in the illustrations). Many clams burrow into sand or mud by means of a powerful fleshy foot, and a few (oysters, mussels, and others) are attached to the bottom. Some scallops can swim by clapping the two shells together. Earth's oldest clam, *Fordilla troyensis*, was described from Lower Cambrian rocks (ca. 517 million years ago) of easternmost New York State.

Class Gastropoda

(gas-*troh*-poe-*dah*; "stomach foot")
Snails live in marine, fresh-water, and land habitats. Except for land slugs and a number of marine forms (nudibranchs), snails generally have a helically coiled (spiraled) dorsal shell. Snail shells (see *Hormotoma* and *Poleumita* in the illustrations) are common in New York State marine rocks.

Class Cephalopoda

(*seff*-ah-low-poe-dah; "head foot")
Cephalopods are marine animals that have the ancestral mollusk foot modified into a ring of tentacles that surrounds the mouth. With a hard jaw (and sometimes a poison sack), suckers or hooks on the tentacles, and an ability to swim by jet propulsion, they are active predators. Although squids and octopi lack external shells, many early cephalopods had straight or planispiral, chambered, external shells (the chambers were filled with nitrogen as in the modern pearly nautilus and assisted in buoyancy control). Two types of cephalpods appear in New York State rocks. Nautiloids (*naw*-till-oidz), close relatives of the modern pearly nautilus, have simple curved walls (called septa) between the chambers. Straight nautiloid shells from Upper Cambrian rocks (ca. 495 million years ago) of Washington County, eastern New York, are among the oldest known on earth. Nautiloids are the most common cephalopod fossils in New York (see *Nephriticeras* in the illustrations). Their descendents, the ammonoids (*am*-uh-noydz), are rarer, first appear in the Middle Devonian (ca. 380 million years ago), have crinkled septa, and became more abundant than nautiloids later in the "Age of Dinosaurs."

Phylum Mollusca?

Tentaculitids

(ten-*takh*-you-lih-*tidz*)
This interesting group of extinct animals are regarded by some paleontologists as mollusks (see *Tentaculites* in the illustrations). They are known from small, cone-like, relatively thick-walled calcareous shells with transverse rings. The small bulb-like end of the cone is likely the larval shell.

Phylum Arthropoda

(*are*-throw-*poe*-dah; "jointed legs")
Arthropods have a relatively hard external skeleton (exoskeleton) with jointed legs and a body that is segmented along its length. Common in marine, fresh-water, and land habitats, this is the largest and most diverse animal phylum. Insects, spiders, scorpions, crabs and lobsters, trilobites, and many other groups are arthropods.

Class Trilobita

(try-low-*bye*-tah; "three lobes")

These extinct marine arthropods are commonly found disarticulated or enrolled. "Trilobite" refers to the longitudinal division of the body into a central and two lateral lobes, and not to the fact that the body is also divisible into a head, body, and tail plate (see *Proetus* and *Phacops* in the illustrations). Only the dorsal exoskeleton of trilobites is typically preserved, because it was mineralized with calcium carbonate. The first preserved legs and gills on the underside of the trilobite body were found in the 1870s in specimens from Trenton Falls, north of Utica, New York.

Class Merostomata

(mare-oh-*stow*-mah-tah; refers to the strong development of walking legs)

This arthropod class includes several near-shore marine groups: the living horsehoe crabs and the extinct "sea scorpions" or eurypterids (you-rip-*tur*-idz; see *Eurypterus* in the illustrations). Eurypterids are locally common in Late Silurian rocks of central New York, and include *Eurypterus remipes*, the New York State Fossil. The largest known arthropod (3.6 m, or 12 feet long) is a giant eurypterid from New York State.

Phylum Echinodermata

(ee-*khy*-noh-*dur*-mah-tah; "spiny skinned")

Echinoderms are an important marine phylum whose members commonly show a five-fold radial symmetry. Starfish, brittle stars, sea cucumbers, crinoids, sand dollars, and sea urchins are echinoderms. A skeleton of loose to tightly interlocked calcareous (limey) plates is secreted within the skin. Embryonic development and blood pigments show that echinoderms are distantly related to other invertebrate phyla, and more closely related to chordates (e.g., vertebrates and relatives).

Class Edrioasteroidea

(*ee*-dree-oh-*ass*-tur-*oi*-dee-ah)

These uncommon forms look like star-fish attached to a cushion-like disc (see *Agelacrinites* in illustrations). Members of this extinct, primitive echinoderm group were sessile or moved slowly on firm or hard sea bottoms. Many are less than 2.5 cm (1 inch) in diameter).

Class "Cystoidea"

(sis-*toy*-dee-ah; "resembling a sack")

"Cystoids" include two small groups of extinct echinoderms with a globular, often egg-shaped theca with arms used in filter feeding, and generally with a stem that attached the animals to the bottom (see *Lepidocrinites* in the illustrations). Blastoids and crinoids (described below) have a more regular arrangement of plates on the theca and calyx (the latter term in crinoids is equivalent to the theca), and lack the pores or slit-like openings that extend through the theca plates in "cystoids." "Cystoids" are most common in Silurian limestones and shales in central New York State, and are now divided into two classes.

Class Blastoidea

(blas-*toy*-dee-ah; "resembling a bud")

Blastoids are similar to crinoids, except that the feeding arms (which are rarely preserved) are smaller and more numerous; the stem is generally much thinner, and the theca has a very uniform five-fold arrangement of plates. Blastoid theca had interlocked plates and are locally common in Middle Devonian rocks of the Finger Lakes region.

Class Crinoidea

(kri-*noy*-dee-ah; "resembling a lily")

Crinoids or "sea-lilies" are very beautiful when a complete calyx with well preserved arms and an attached stem is found. The calyx plate arrangement is regular and complex, and the arms are used in filter feeding. Crinoids have segmented stems that consist of circular or star-shaped columnals with a central hole. These columnals are very abundant in some limestones. Root-like holdfasts anchored the animals to the sea bottom (see "Crinoid" in the illustrations). Species that live today include stalked forms in the deep sea and shallower-water forms that shed the stalk as adults and walk around on their arms.

Class Asteroidea

(ass-*tur*-oi-dee-ah; "resembling a star")

Ancient starfish were similar to those living today, both in structure and habits (see *Devonaster* in the illustrations). They have always been predators on slow-moving marine animals. Because the plates in a starfish skeleton are scattered after death by decay and scavengers, rapid burial by sediment during storm events explains the preservation of complete specimens.

Phylum Hemichordata

(hem-*ee*-kore-*dah*-tah; "half chordates")

Hemichordates include several small groups of living marine animals that share some features with vertebrates (e.g., a dorsal nerve cord). The extinct graptolites are thought by many paleontologists to be hemichordates.

Class Graptolithina

(*grap*-toe-*lih*-thee-nah; "rock writing forms")

Graptolite (*grap*-toe-lite) fossils often resemble black pencil markings on broken rock surfaces. These marine animals were colonial and produced tough, flexible, protein-walled exoskeletons. Some colonies were fan-like and resembled the fan-like colonies of calcareous bryozoans. Most graptolite colonies look like saw blades, with each tiny notch occupied by a living animal during life (see the illustrations). New York graptolites are a standard for world-wide correlation of Ordovician and Silurian rocks.

Phylum Chordata

(*kore*-dah-tah; refers to the dorsal stiffening rod [notochord] and dorsal nerve cord in vertebrates and relatives)

Class Conodonta
(koh-noh-*don*-tah; "cone teeth")
These extinct animals are the most important group for global correlation of Upper Cambrian through Triassic marine rocks (ca. 495–195 million years ago). Their small tooth like jaw parts (conodont elements) are composed of phosphatic minerals and can be dissolved out of limestone samples. Rare finds of fossil conodonts with soft parts preserved show that that had eyes and the V-shaped lateral muscles that fish have, and are closely related to vertebrates. Conodonts had the ecologic role of small predatory fish before the diversification of jawed fish.

HISTORY OF LIFE IN NEW YORK

The vastness of geologic time is difficult to comprehend. Paleontologists come to appreciate this "deep time" because they are particularly interested in reconstructing the succession and evolution of life and the ancient environments recorded by fossil occurrences. Paleontologists' understanding of the succession of life led to development of the geologic time scale and its divisions (see the illustration of the time scale). Most divisions of the geologic time scale were defined on the basis of characteristic fossil assemblages and named in the 19th century. By the mid-1800s, a geologist in any area of the world could determine and communicate the relative age of the rocks that he/she was working on. This was done by reviewing the fossils that were found and comparing them to those known in the areas where the time scale was developed.

The geological time scale is hierarchical. The longest time divisions, the eras, were named very early and describe the types of organisms that lived during these intervals. The eras over the last 543 million year interval when fossils become abundant include the Paleozoic, Mesozoic, and Cenozoic Eras. These designations mean, respectively, "ancient life," "middle life" (often called the "Age of Dinosaurs"), and "recent life" ("Age of Mammals"). Eras are divided into geologic periods, which were generally named for areas where the rocks and their fossils were believed to be most representative. For those geologic periods represented by New York rocks, the Cambrian, Ordovician, Silurian, and Devonian were named for fossiliferous areas in southern Britain, while the Carboniferous ("carbon-bearing") was named for its abundant coal in western Europe. "Age of Dinosaurs" rock in southeastern New York is referable to the Triassic (refers to the characteristic three [tri-] vertical divisions of these rocks in western Europe), Jurassic (named for the Jura Alps in France), and Cretaceous (Latin for the "chalk-bearing" Cliffs of Dover).

Geologic periods are further divided into epochs, and epochs into ages. After a long interval following the Cretaceous Period in which no fossils or rocks are known in New York, abundant fossils are known from Pleistocene Epoch ("Ice Age") sediments across New York. The Pleistocene and younger Holocene (or Recent, beginning 10,000 years ago and continuing today) Epochs comprise the Quaternary Period (an old name meaning "fourth").

The names of eras, periods, epochs, and ages are constantly used by geologists and paleontologists to communicate the age of rocks, fossils, and geologic events. They provide a basic vocabulary for relative time, similar to that given to successive material cultures unearthed by archeologists (e.g., Paleolithic, Neolithic, Chalcolithic, etc.). The student of paleontology has to commit these relative time terms to memory. The absolute ages of fossils or rocks can sometimes be determined by techniques based on radiogenic decay. However, a paleontologist will regularly refer to the relative age of fossils (say, "early Middle Cambrian"), rather than give an absolute age ("about 505 million years"), because precise absolute ages are usually unknown for most fossil horizons.

Precambrian Eon

Little of this most ancient interval of "deep time" is known in New York State. The Precambrian extends from the origin of the earth and solar system 4.5 billion years ago to the beginning of the Cambrian Period and the regular appearance of fossiliferous rocks 543 million years ago. During the Precambrian, the basement of New York, which is uplifted in the Adirondack Mountains and Hudson High-lands, was deposited as sedimentary rocks and volcanic deposits. These ca. 1.3 billion year-old rocks were involved in a major mountain building event 1.1 billion years ago that involved collision of ancestral North America with a continent called "Amazonia." Remnants of the old Amazon continent were recognized in early 2002 to compose the bedrock of the Westchester County–Manhattan region. Deep burial, metamorphism, and melting by great heat and pressure destroyed almost all fossils in the basement of New York. In the 1980s, work by N.Y.S. Museum scientists led to discovery of large, layered cyanobacterial ("blue-green algal") build-ups in St. Lawrence County. These remains, the oldest fossils in the eastern U.S., reflect the "simple," unicellular, bacterial-grade of life 1.3 billion years ago.

Cambrian Period

The earliest period in which fossils first become reasonably common is the Cambrian (*cam*-bree-en), which began 543 million years ago. Many "pages" of the Cambrian chapter of the book of "deep time" are missing in New York. Early Cambrian rocks occur in the Taconic slate belt east of the Hudson River. Rare and fragmentary trilobites, snails, worm tubes, brachiopods, and trace fossils are found, particularly in thin limestone beds in these very deep-water (continental slope) rocks. These are the oldest animal fossils in the Middle Atlantic States. Whereas Early Cambrian shallow seas were restricted to southern Dutchess County, sea-level rise drove Late Cambrian shallow seas across New York. Upper Cambrian rocks rim the

GEOLOGIC TIME SCALE

EON	ERA	PERIOD	EPOCH	Millions of Years	LIFE RECORD	RECORD IN NEW YORK: Fossil – Plants	Fossil – Invertebrates	Fossil – Vertebrates	Rock
PHANEROZOIC	CENOZOIC	Quaternary	Holocene	0	Humans, mastodons, mammoths, condors				
			Pleistocene	2					
		Tertiary	Pliocene	7	Large carnivores				
			Miocene	26	Abundant grazing animals				
			Oligocene	37 38	Large running mammals				
			Eocene	53 54	Many modern types of mammals				
			Paleocene	65					
	MESOZOIC	Cretaceous	Late	100	Dinosaur extinction Earliest placental mammals Earliest flowering plants Climax of dinosaurs and ammonites Great decline of brachiopods Great development of bony fish				
			Early	144					
		Jurassic	Late	162	Earliest birds Abundant dinosaurs and ammonites				
			Middle	172					
			Early	201					
		Triassic	Late	205	Earliest dinosaurs and mammals Abundant cyads and conifers				
			Middle	215					
			Early	245					
	PALEOZOIC	Permian	Late		Extinction of many kinds of marine animals, including trilobites				
			Early	268	Appearence of mammal-like reptiles				
		Carboniferous	Pennsylvanian		Earliest reptiles. Great coal-forming forests				conglomerate
			Mississippian	362	Abundant sharks and amphibians Large and numerous scale trees and seed ferns				
		Devonian	Late		Earliest amphibians and ammonoids				
			Middle						
			Early	418	Extinction of armored fish, other fish abundant				
		Silurian	Late		Earliest land plants and insects Peak development of eurypterids Earliest fishes in New York				
			Early	430	Great diversity of echinoderms and ostracodes				sandstone
		Ordovician	Late		Invertebrates dominant-coelenterates, mollusks and arthropods become abundant; echinoderms diversify; earliest coral reefs in early Middle Ordovician				shale limestone
			Early	489	Graptolites very abundant				dolostone
		Cambrian	Late	500	Earliest fish				
			Middle	510					
			Early	543	Trilobites appear Appearence of animals capable of burrowing deeply into sediments				
Precambrian Eon					Appearence of soft-bodied, many-celled animals Single-celled algae and stomatolites				gneiss
ORIGIN OF EARTH				4500	Fossils extremely rare, primitive aquatic bacteria				

J. Skiba New York State Museum

Some Fossils of New York

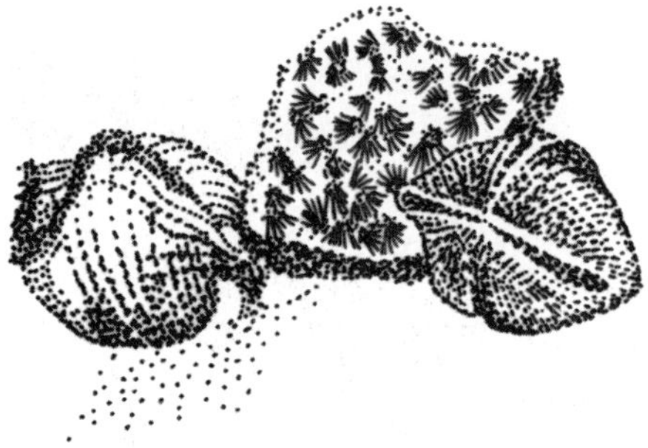

BRACHIOPOD *("Spirifer")*
Devonian

SPONGE *(Hydnoceras)*
Devonian

MOSS ANIMAL (Bryozoan)
Devonian

STALKED SAC–ANIMAL, "CYSTOID"
(Lepidocrinites)
Devonian

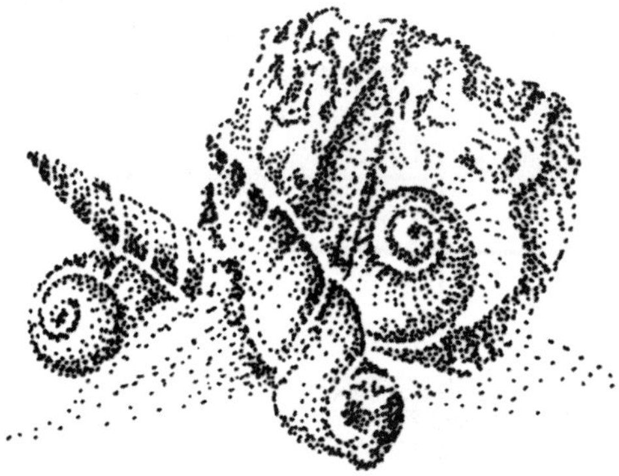

SEA SNAIL
(High—*Hormotoma)* (Low—*Poleumita)*
Silurian

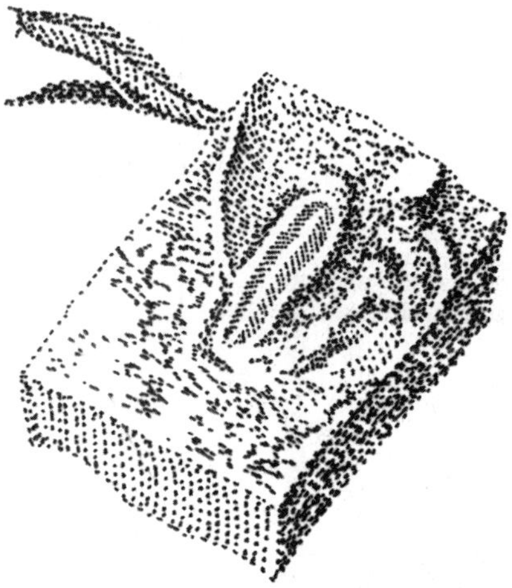

SEA WORM *(Protonympha)*
Devonian

MOLLUSK-LIKE ANIMAL *(Tentaculites)*
Devonian

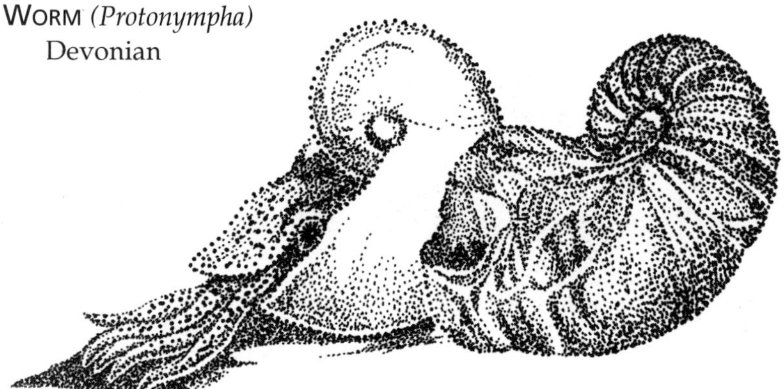

CEPHALOPOD *(Nephriticeras)*
Devonian

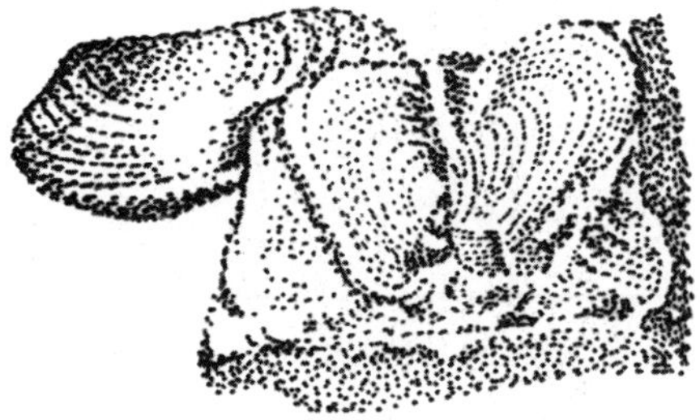

CLAM *(Glyptodesma)*
Devonian

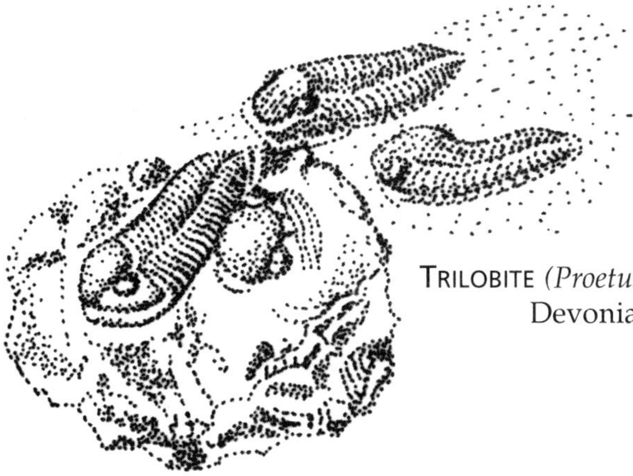

Trilobite *(Proetus, Phacops)*
Devonian

Sea Scorpion *(Eurypterid-Eurypterus)*
Silurian

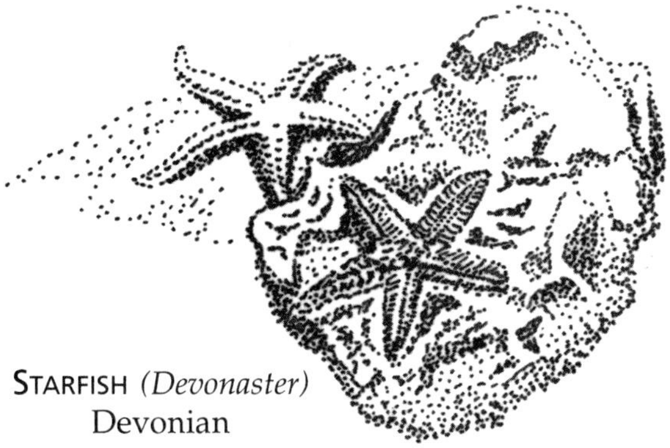

Starfish *(Devonaster)*
Devonian

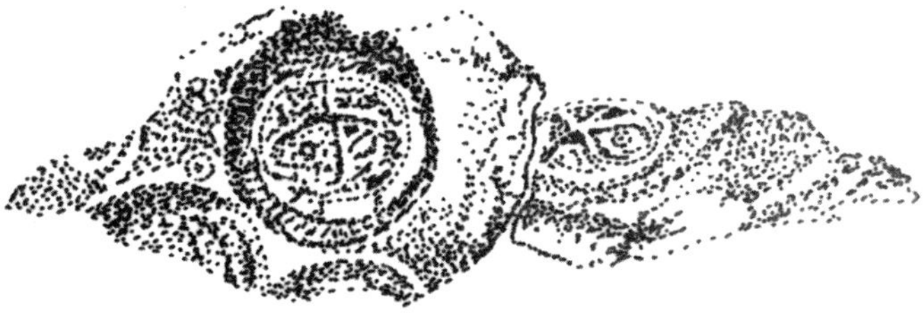

Star-Disc Animal—Edrioasteroid
(Agelacrinites)
Devonian

Saw-Edge Animal (Graptolite)
Ordovician

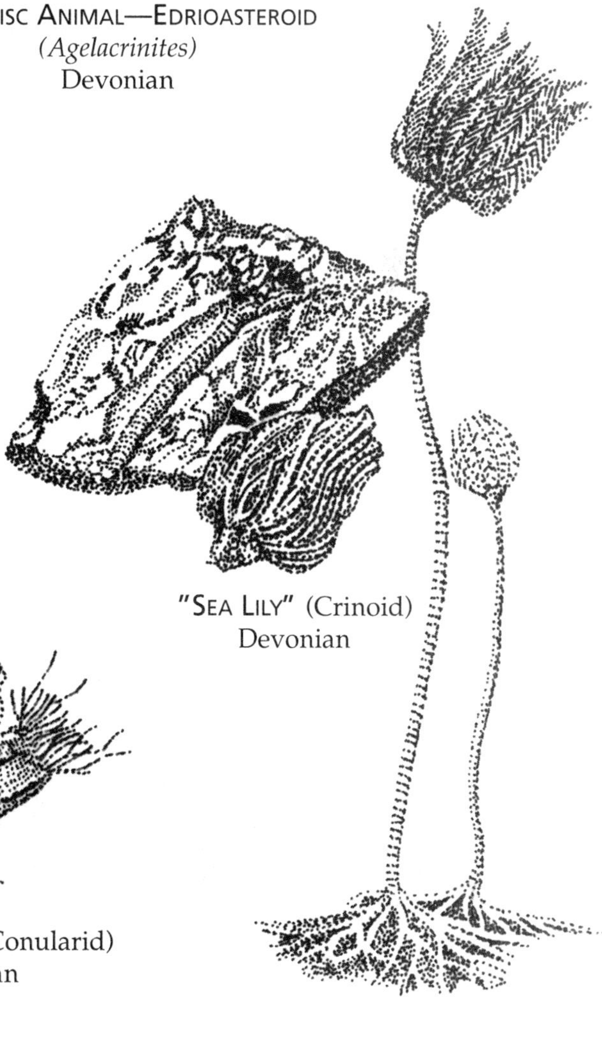

"Sea Lily" (Crinoid)
Devonian

Coral *(Zaphrentis)*
Devonian

Cone-Hydras (Conularid)
Devonian

Adirondack Mountains, and crop out in the Mohawk River valley and from Washington to Dutchess County. New creatures, such as nautiloid cephalopods and other mollusks appeared in the Late Cambrian. Numerous types of trilobites and trace fossils of soft-bodied organisms appear in the quartz sandstones deposited near the beach, like those exposed in Ausable Chasm, and in deeper-water lime mudstones. Cyanobacterial reefs, as those at Lester Park and Petrified Gardens near Saratoga, are preserved as stromatolites ("layered rocks").

Ordovician Period

The Ordovician (*or*-doe-*vish*-ee-an) Period began 489 million years ago and lasted 50 million years. The Ordovician of New York is a global reference standard for this period. Early and Middle Ordovician gastropods, nautiloids, and ostracodes are particularly diverse in the shallow-water, tropical carbonates widespread in the state. By comparison, trilobites are not particularly abundant or diverse in these shallow-water rocks. At the same time in the deep waters of the Taconic slate belt, graptolite-rich muds accumulated. Trilobites have their highest abundance and diversity in the Late Ordovician with regional deepening and appearance of muddy bottoms in the Late Ordovician. This deepening was related to collision of a Japan-like volcanic island belt with eastern North America and mountain building along the edge of the continent. This collision pushed the Taconic slate belt into eastern New York. The combined load of the slate belt and island arc actually pushed all of New York down under its weight in the Late Ordovician.

The last Ordovician tropical, shallow-water limestone deposition in New York took place before the mountain building caused by Late Ordovician collision with the volcanic island arc. These limestones, which are seen near Plattsburgh and Chazy, New York, show earth's oldest stromatoporoid and coral reefs. Reefs have been important through geologic time because they supply a variety of environments that other animals and plants live in or on. This is recorded in the northern Lake Champlain lowlands by the abundant snails, nautiloids, bryozoans, echinoderms (notably "cystoids"), and trilobites that occur around the Chazy reefs.

Even richer fossil assemblages are found in somewhat younger limestones of the Late Ordovician Trenton Group in the Black River, Mohawk, and Lake Champlain valleys. These limestones are not the shallow-water limestones of the Early and Middle Ordovician, but are deeper-water and often clay-rich. By comparison, mollusks, although present (e.g., clams, snails, and nautiloids), are less abundant and trilobites far more diverse in Trenton limestones. Locally, brachiopods compose whole limestone beds low in the Trenton Group to a near exclusion of trilobites. Crinoids, starfish, and bryozoans first become relatively abundant in New York in Trenton limestones. Trenton limestone deposition ended with further depression of eastern New York and rapid deposition of mud eroded from the Taconic mountains. Low-oxygen waters overlay the lightless, deep mud bottoms of this "Utica Sea," and few bottom-dwelling animals could live on its bottom. Graptolites colonies, perhaps reflecting a floating life habit or, more likely, an adaptation to low-oxygen environments, settled into this mud, and are common fossils.

The Late Ordovician ended with deposition of near-shore to sub-aerial (i.e., deposited on land) red shales, beach sandstones, and near-shore marine gray shales that comprise the deposits of the huge Queenston delta. The Queenston delta extends from central and western New York (where its rocks are seen at river-level below Niagara Falls) and across Ontario and Ohio. The shallow-water environments of the Queenston delta probably reflect the filling of the sea in eastern North America with sediment eroded from the Taconic uplands. In addition, global sea-level fall with the major end-Ordovician glaciation of North Africa and South America (which were at the South Pole) is likely part of the explanation for the end-Ordovician shallowing of the seas in New York. It should be noted that continental glaciation causes sea-level fall, as water evaporated from the ocean is deposited on land as snow and ice. Fossils are not abundant in Queenston sediments, and include mollusks (as usual, clams and snails are more abundant in these shallow facies) and trace fossils. Queenston delta rocks are particularly interesting, however, as they yield spores produced by mosses—earth's oldest land plants.

Silurian Period

Melting of the North African and South American ice caps caused global sea-level rise at the beginning of the Silurian (sigh-*lure*-ee-an) Period. Fossiliferous Silurian (438–418 million years ago) rocks are largely limited to western New York. Eastern New York remained uplifted after Taconic mountain building and was not covered by seas until later in the period. The earliest Silurian rocks, best seen low in Niagara gorge, include beach, near-shore, and stream deposits that have the same types of fossils (clams, snails, ostracodes) that characterize near-shore Ordovician deposits. A long, segmented-appearing horizontal trace fossil called *Arthrophycus* appears in the Lower Silurian and ranges from western New York to Alabama.

Deeper-sea conditions and more open-marine fossils (brachiopods, ostracodes, some trilobites) are recorded by overlying Lower Silurian limestones and shales in western New York. However, open-marine conditions appear only higher in the Lower Silurian where they are reflected by the over 200 species of the Rochester Shale (trilobites, corals, bryozoans, cephalopods, crinoids, etc.). The massive limestones and dolostones of the Lockport Group above the Rochester Shale form the caprock of Niagara Falls and record shallow tropical sea conditions with coral reefs.

Slow sea-level rise in the Late Silurian after Lockport Group deposition allowed the shoreline to reach the Hudson River area in the latest Silurian. Late Silurian seas

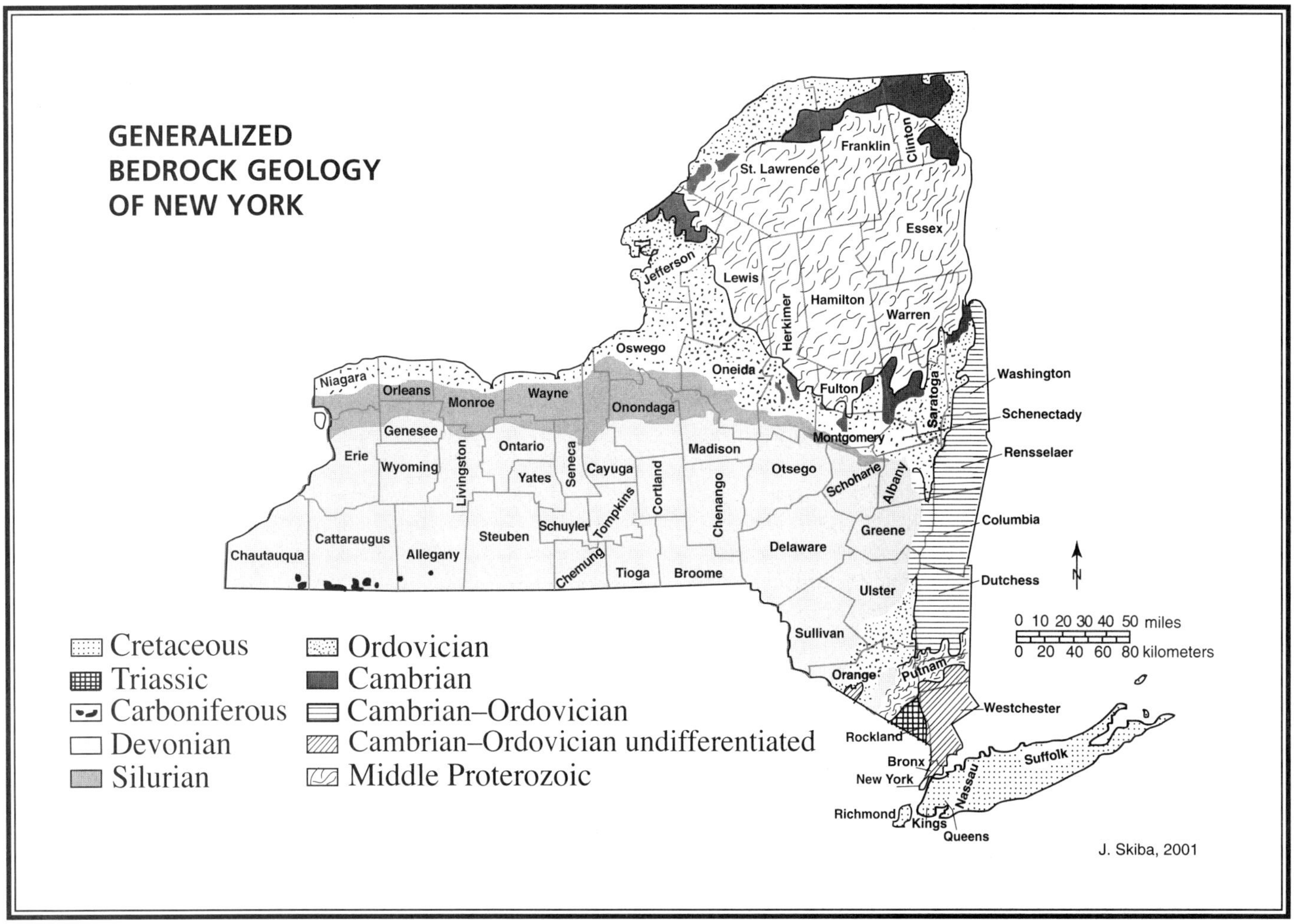

across New York and eastern North America were generally very salty, and largely devoid of organisms. Intervals of more normal marine salinity do occur in Late Silurian rocks in New York, and are recorded by thin intervals with brachiopods, ostracodes, and eurypterids. Accompanying these near-shore marine animal assemblages are specimens of scorpions and fragments of *"Cooksonia,"* earth's oldest vascular plant; these rare fossils were washed in from land environments.

Devonian Period

With half of New York covered by fossiliferous rocks of the Devonian (deh-*vone*-ee-an) Period (418–362 million years ago), New York State has long been a standard for world-wide correlation of Devonian rocks. These rocks dominate New York east of the Hudson River and south of the Mohawk River–Lake Ontario lowlands. The oldest Devonian rocks are fossiliferous limestones and shales of the Helderberg Mountains in eastern New York. Brachiopods, echinoderms, and bryozoans are locally abundant in the Helderberg Group Limestones. A low-diversity fauna of tentaculitids, snails, ostracodes, and clams appears in the shallow-marine Manlius Limestone at the base of the group. The Helderberg Group has the deepest marine-shelf rocks in its middle part, where calcareous shales of the New Scotland Formation resemble the older Rochester Shale, and yield a comparable number (250 species) of fossils.

The younger Middle Devonian Onondaga Limestone and the succeeding Hamilton Group of formations are known worldwide for their diverse marine faunas. The Onondaga, with its coral reefs, is the last tropical marine limestone deposit in New York. Onondaga limestone deposition was snuffed out with the erosion and transport of enormous amounts of mud and sand from uplifted areas in the interior of the Appalachians. This uplift is the second interval of Appalachian mountain building, the Acadian orogeny ("mountain building event"). This orogeny resulted from collision of a New Zealand-sized microcontinent termed "Avalon" with the eastern margin of ancestral North America. The muddy deposits of the Hamilton Group, which are well exposed in the Finger Lakes and extend west to the Lake Erie Bluffs, change eastward into sandstones in the high peaks region of the Catskill Mountains. The near-shore and subaerial deposits of the Catskills preserve earth's oldest fossil forests in Schoharie County, eastern New York. Clams and fish from ancient stream deposits and land-living arthropods (large spider-like forms, centipedes, and microscopic mites and flightless insects) in Catskill rocks record the oldest fresh-water and land-animal communities. It is important to note that fallen leaves and wood provide nutrients and food that are the basis of fresh-water and land communities. Advanced air-breathing fish with anterior and posterior fins with bones that allowed walking are known in Québec and Pennsylvania rocks identical in age to those of the Catskill Mountains. These ancestors to the amphibians will likely be found in New York.

Middle Devonian rocks of New York record the decline of trilobites, perhaps as a result of the rise of predatory jawed fish and increased competition from advanced arthropods; only a few trilobite groups survived the Devonian. More important for global correlation of New York Devonian marine rocks are conodonts and ammonoid cephalopods that appeared in the Middle Devonian and diversified through the Late Devonian. Graptolites are rare in Lower Devonian rocks in New York, and did not survive the Devonian. Brachiopods remained abundant in Devonian rocks in New York, and brachiopod families that dominated the succeeding Carboniferous Period appeared in the Middle Devonian. However, an impressive diversification of clams and the shift of diverse clam communities from shallow to deeper marine habitats heralded the gradual replacement of brachiopod-dominated bottom faunas by clam-dominated assemblages. As described above, the "drama" of life shifted to the land in the Middle Devonian with the rise of forests and fresh-water communities. Although sharks appeared in marine settings in the Middle Devonian (their teeth and dermal spines appear in the Onondaga Limestone), the diversification of advanced bony fish and the origin of amphibians in the Late Devonian took place in fresh water.

The early Late Devonian featured deposition of deep-water, black, poorly fossiliferous shale across western and central New York. These deposits resemble the Utica Shale deposited during the much older Taconic orogeny (described above). However, sediment eroded from the Appalachians progressively filled this deep-water belt so that the latest Devonian had shallow-water clam and brachiopod faunas in Allegany and Chatauqua Counties.

Carboniferous Period

Carboniferous Period (362–286 m.y.a.) rocks are limited to a few hill tops in the Allegany State Park area of southwestern New York. They represent only two parts of the Carboniferous. The earliest Carboniferous rocks include marine shale and sandstone that resemble latest Devonian rocks, and have clam-dominated faunas with transported land plant debris. The overlying rock near Olean Rock City is a non-fossiliferous, massive conglomerate that extends into Pennsylvania, where overlying rocks have Middle Carboniferous plants and coal.

Mesozoic Era

Mesozoic (245–65 m.y.a.), or "Age of Dinosaurs" rocks are found in southeast New York. Dinosaur tracks were found in 1972 in ancient lake-margin deposits at Blauvelt, Rockland County. This site has not been completely described in the scientific literature, but is known to have dinosaur teeth and tree stumps in life position. Latest Triassic to earliest Jurassic sites (ca. 220–190 m.y.a.) with dinosaur tracks along the Connecticut River in Massachusetts and Connecticut are the same age as the Blauvelt locality, and have fresh-water fish, clam, and arthropod fossils.

Late Mesozoic (middle Cretaceous Period, ca. 100 m.y.a.) sandstone and shale with leaves and branches of conifers and early flowering trees (magnolias, fig-like forms) were once accessible in quarries in Staten Island and western Long Island. Land organisms (as insects) do not seem to have been recorded in these deposits. Similarly, the vertebrates (mosasaurs [marine crocodile-like forms] and birds) known in the Cretaceous of New Jersey are not reported in New York.

Pleistocene Epoch

A gulf of a hundred million years separates the Cretaceous rocks and fossils of southeast New York from Pleistocene (2 million years ago to 10,000 years ago; the "Ice Age") sediments that cover New York. Plants and animals (likely including humans) appeared in New York with the melting and retreat of continental ice sheets that began ca. 20,000 years ago. Early cold-adapted animals included elephant-like mastodons and rare mammoths, pig-like peccaries, caribou, deer, moose, fox, and birds (including a condor found in western New York). These vertebrates occur with water- and land-plant remains in stream and bog deposits.

Marine Pleistocene fossils are found in northern New York. Because the earth surface was depressed more than 550 feet under the weight of the continental ice sheet, its melting led to an initial incursion of marine waters into Lake Champlain and the upper St. Lawrence River. Clams (including oysters), ostracodes, and foraminiferans (single-celled animals) have long been known from this region, and indicate cold, brackish waters. A slow uplift of northern New York from Lake Champlain to the Niagara River frontier continues as the last effects of downwarping under the weight of the continental ice sheets are overcome.

FOSSILS: CONSERVATION AND COLLECTING

Fossils should not be regarded as curiosities to be bought and sold as attractive knick knacks. They should be appreciated as a limited resource that is the record of the parade of life on earth and the key to helping reconstruct it. Rocks with their entombed fossils mirror an orderly succession of life and events. This intrinsic value of fossils means that we should do our best to conserve them. There are a limited number of exposures of fossil-bearing rocks along streams, rivers, and lake shores and in artificial cuts along roads and railroads.

1

However, there is large population of potential collectors. Commercial collectors and visits by large groups can completely strip localities of fossils. This means that a site can be easily destroyed for later use in field trips by public school through university groups or in professional research. For these reasons, fossil collecting should be limited to loose material that is no longer in context and to places (excavations and quarries) where the rocks and fossils are already in process of removal. Just as bird watching and photography replaced destructive specimen hunting, fossil "hunting" should now largely involve looking, recording, photographing, and leaving only your footprints. Permission must be asked if collecting is done on private lands. In addition, removal of fossils or rocks from state parks, forests, and canals and from road cuts along state and federal roads is not allowed without special permits.

However, the responsible collecting of fossils, with the self-imposed limitations suggested above, provides the most direct introduction to an understanding of "deep time." Paleontology is a hands-on, empirical science that is decidedly analog, and has an important field component. It cannot be practiced, advanced significantly (the important data either hasn't been completely collected, or the available data has frequently been collected with fundamental biases and needs field checking), or enjoyed through work limited to a computer. Collecting and field experience are vital. The collector should have a reason for collecting and should focus collecting on reasons other than getting "more stuff" to fill his/her shelves.

Collecting fossils does not involve "luck." The best preparation for finding them is knowing where to look, which comes from understanding aspects of the geology of an area. The professional paleontologist knows that the best field work is done in the library by reviewing published geologic maps, scientific reports, and field trip guides. As noted above, the very ancient and metamorphosed rocks of the Adirondacks, Hudson Highlands, and the younger, but metamorphosed, rocks of the Westchester-New York City area and eastern ridges of the Taconic Mountains will be a paleontological desert. However, almost any other bed rock area in the state will yield fossils.

The primary equipment for field collecting of fossils includes sharp eyes, patience, a few hand tools, and, most importantly, a field notebook (1). The field notebook is emphasized because a fossil without context is just a "thing" that communicates little about its role and significance in the book of "deep time." Professional paleontologists carefully record the precise geographic locality for every fossil they collect for study (i.e., country, state, county, township, location on a U.S. Geological Survey topographic sheet, GPS coordinates), as well as the geologic context (i.e., location in a section of rock, type of rock, etc.).

At some localities, fossils may be found loose and weathered from rock or in sand or loosely cemented rocks. In New York State, this is the exception, and hand tools must be used to collect them from hard rock. A stone mason's hammer with a chisel edge (2) and a three-pound sledge hammer (3) are most useful. Specially hardened cold chisels (4, 5, 6, 7) help in breaking rock or splitting it along bedding planes. Chisels that are at least one inch in diameter at their head reduce hand injuries that result from missed hammer strikes on thin chisels. Injuries from flying rock and steel are reduced if gloves are worn. Eye protection is mandatory (8) if safety glasses are not regularly worn. Quarry owners will often mandate a hard hat if they give permission to enter their property, and the collector should bring his/hers. A small magnifying hand lens (9) (no more than 7x power) is a must to see small fossils and the fine structures of larger ones. Fossils should be wrapped in the field with newspapers, tissue paper, or foil (10) to preclude damage. Small bags (11) are good for small fossils; canvas bags with a drawstring (12) are necessary for larger slabs. Delicate fossils

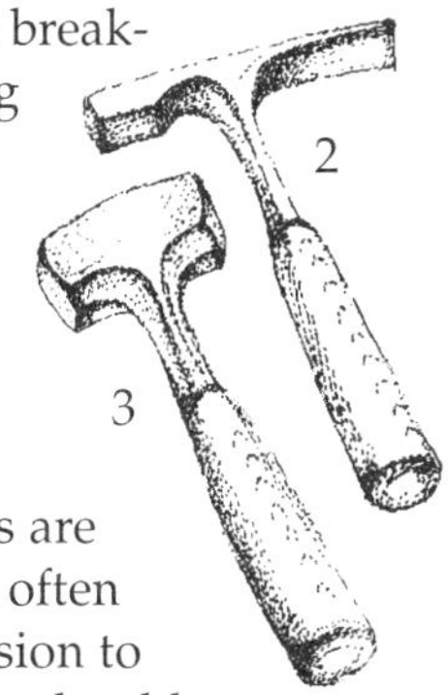

2 3

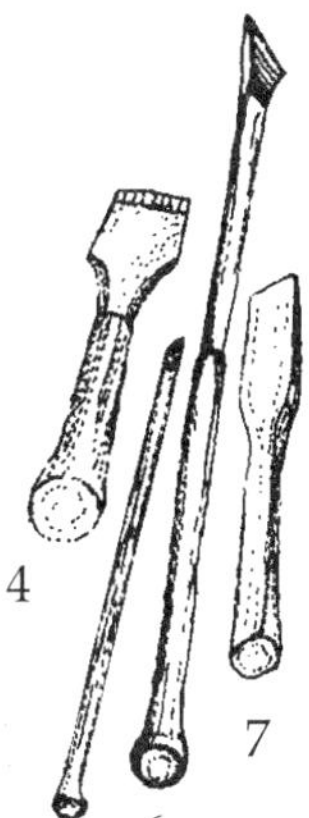

4 5 6 7

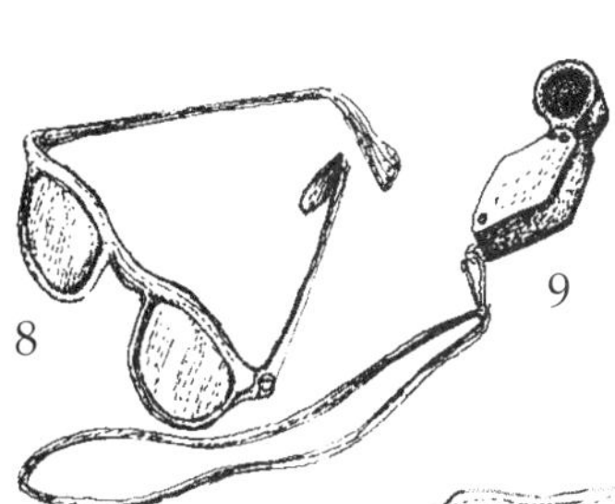

8 9

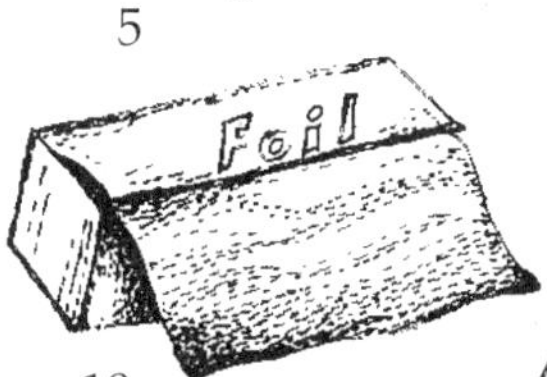

10

11

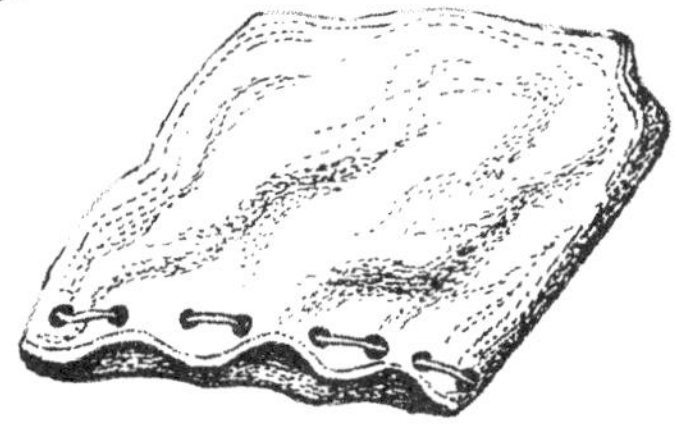

12

should be wrapped and then packed in cardboard or plastic containers (13).

Fossils must be labeled (14) in the field (the memory is too fallible!). I assign a field number, and write it on the specimen or its bag. This field number is a letter and number combination that records the section/locality name and the sample's position in the section (e.g., a designation like "On-13.5" means "Onesquethaw Creek," 13.5 meters from base of measured section). I also always mark an "up arrow" in magic marker on each specimen. The "up arrow" is perpendicular to bedding and points up in the direction of overlying, younger strata. The "up arrow" may be important in later study of the fossils, for example complete trilobites are generally found upside-down; they were almost always overturned by storms or under-water sediment flows and buried alive by sediment. The more complete notes in the field book should include an exact drawing of the carefully measured rock section that records the types of rocks through the section, the thickness of all rock units (always measure in metric units), the exact location of the section, the name of the rock formation (if known), name of fossils (if known) at each level through the section, and the name of the collector and date of work. A camera helps record the geologic context of fossils before removal from the rock. If fossils are broken during recovery, the resultant jig-saw puzzle is more easily solved if a photo was taken. Duco household cement or a similar plastic cement is an indispensable repair aid. Finally, a sturdy backpack (15) helps carry specimens. Shovels, trowels, and whisk brooms may be used to help uncover bedding surfaces or in sampling soft sediments. Bones and teeth from Pleistocene vertebrates are relatively rare, fragile, and important, and should be left in place; professionals should be brought in to collect them. If a Pleistocene bone is collected, it should not be allowed to dry out and fall apart before being brought to a professional.

13

14

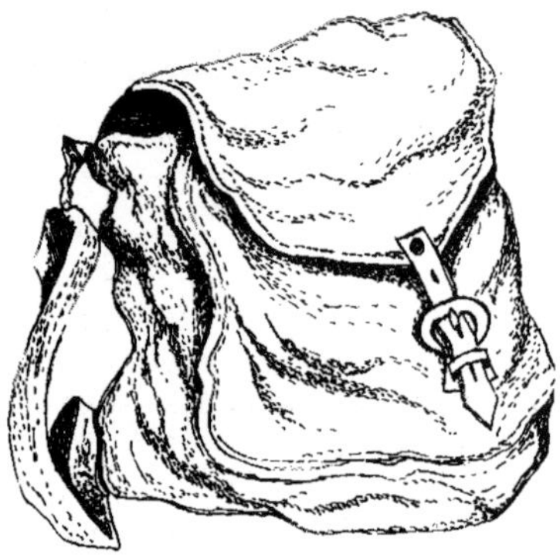

15

The nature of the fossil and surrounding rock matrix determines how the specimen is prepared after being brought in from the field. Some matrix can be removed with a hammer and chisel or with an electric masonry saw. Brushes (16) with plastic or metal bristles can remove soft matrix. Mounted needles, called teasing needles (17), or dental picks are time-consuming, but useful in matrix removal. A small electric engraving tool, called a Vibra-tool (18), especially if fitted with a hard needle and used under a microscope, can reveal the most detailed features on complex forms like trilobites. The Vibra-tool takes practice, but is used by professional preparators. Acids help recover fossils with a chemical composition that is different from the matrix. This is particularly the case in recovering silicified fossils or phosphatic fossils (conodont elements, some brachiopods, fish teeth and bones) from limestone (calcium carbonate) by immersion in dilute acetic or formic acid, which dissolves the matrix. Hydrochloric acid (also called muriatic acid by plumbers) is a much stronger acid, and can only be used to recover silicified remains from limestones.

Detailed techniques are required for the identification of most fossil sponges and stromatoporoids, corals, bryozoans, and cephalopods. As these fossils are distinguished at the genus- and species-level by internal morphologic details, polished surfaces and thin-sections must be made. A polished flat surface is made by cutting the specimen on rock or masonry saws, which use tiny industrial diamonds embedded in steel to cut hard, brittle materials). Polishing is done by grinding the cut surface with progressively finer carborundum powders that have been spread on glass plates and moistened. A final polishing is done with jeweler's rouge or tin oxide.

Care must be taken that the glass plates are thoroughly washed before the next grade of grinding powder is used, or a separate glass plate is used for each grade of grinding powder. Contamination can cause undesired scratches. A polished look without polishing can be achieved by coating the section with colorless plastic spray (Krylon brand). Prior to final polishing, it may be helpful to make a transparent peel. This is made by slightly etching the ground

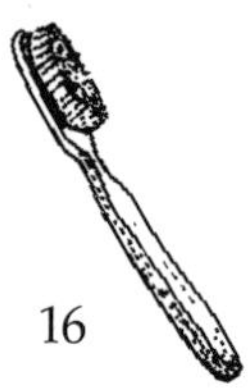

16

17

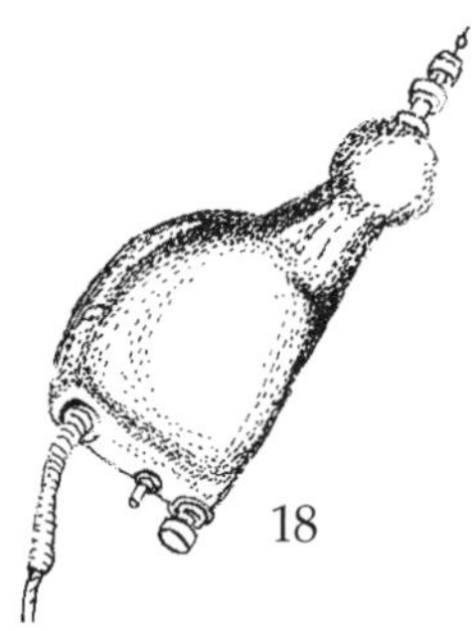

18

surface with very dilute hydrochloric acid, then washing the acid off, and drying the etched specimen. The etched specimen is sprayed with acetone (this is explosive around sparks or open flames!), and a small piece of acetate paper is lain on the acetone-wetted surface. With evaporation, the acetate paper flows into the etched surface and makes a mold of the fine details. When detached, the acetate paper can be examined with a microscope or used as a negative to make a photographic print. To make thin sections, a polished thin slab is glued to a glass slide with Canada Balsam, bioplastic, or a similar adhesive. The section is then ground down until it is so thin (0.03 mm is the standard) that light passes through it, and structural details are visible under a microscope.

If a fossil occurs as a mold (i.e., the shell is dissolved away to leave a hollow in the rock), it may be advantageous to make a cast that reproduces the shell. This is done by pouring liquid dental plastic into the hollow, allowing it to set, and then pulling it out of the mold.

Photography helps in the study of fossils by revealing and preserving details not visible under ordinary lighting. Proper exposure and lighting in light photography are gained through experience. In general, flat fossils should have low incident light; rough or high relief ones should have high incident light. Usually two lights are used, the stronger coming from the upper left. It is best (actually "required" for good results) to coat specimens photographed by light photography with a thin dusting of ammonium chloride. This dusting helps emphasize surface details, gives a uniform level of reflectance to the specimen, and is easily washed off with water. This dusting is produced by blowing air slowly through a tube of ammonium chloride heated by a bunsen burner and allowing the ammonium chloride "smoke" to condense on the fossil. Completely flat fossils, as many graptolites in shale, show enhanced color differences with the rock matrix if shot under water or glycerine. Infrared film is also useful in photography of completely flat fossils under liquids. Finally, scanning electron microscopy allows recording of features on small fossils and provides the greatest depth of field.

Useful collections are based on the orderly storage of specimens. They should be properly labeled and stored in as dust-free a place as possible. Cardboard trays, floored with archival polyethelene batting, and separate inked labels in shallow drawers are the usual method. If possible, all paper materials should be acid-free and of archival quality. Cotton should not be used, as it attracts water from humid air and leads to breakdown of paper labels and fossils. The collection should be kept in a dry storage area—high and fluctuating temperature and humidity cause "pyrite disease" in many rocks. Pyrite disease seems to be mediated by bacteria in humid environments. Pyrite, which is finely distributed in black shale, many sandstones, and the pores in many fossil bones, breaks down under pyrite disease to yield sulphuric acid. This acid breaks the rock apart and corrodes labels and boxes. Pyrite disease can destroy fossils within five years in humid buildings. Expensive, plastic transparent boxes must not be used to store fossils. Plastics always exude volatile resins that tend to corrode fossils and paper labels.

The purchase of fossils for a collection is discouraged. Purchase encourages professional collecting and, frequently, the destruction of unique localities that may be located on public lands. Many fossils supposedly collected in developing or "third world" countries are skillful frauds. For example, many of the complete Cambrian trilobites from Morocco are composites of several trilobite species glued together with automobile fiberglassing techniques. Many specimens are made only of a mixture of ground rock and fiberglass resin. Over the years, I have had visitors to the State Museum show me "fossil birds" and tiny "complete dinosaur skeletons" from China that were made by acid etching of marine limestone slabs. The best fraud I have seen involved Indonesian chert specimens with "snakes" and other vertebrates in living color inside them! The almost transparent chert pieces had been cut in half, and had the shape of half an animal cut into each half. The hollows were filled with colored sand introduced through a tiny hole after the chert slices were glued together.

Acknowledgments

This brochure is an update of Dr. Donald W. Fisher's (N.Y.S. Paleontologist, retired) *Mirror to the Past,* which originally appeared in a 1959 issue of *The Conservationist*. The figures of fossil organisms were originally drawn by Wayne Trimm.

NOTES

93-83617 93-101 CDC